EXTREME ANIMALS

ACTIVITY BOOK

Bear Grylls

Bear Grylls

On my adventures, I've had to put all my skills and knowledge to the test to survive in some of the harshest and most dangerous habitats. It's a tough world out there in the wild, and the different ways some animals have adapted to survive in extreme conditions is absolutely mind-blowing. From animals with super senses and super speed, to the biggest giants on the planet, there is so much to learn about survival from these incredible creatures.

In this activity book, you will learn all about some of the world's most amazing extreme animals, and pick up some incredible facts along the way.

Bear

How to use this book

There are lots of amazing activities to tackle in this book, from mazes and word searches to dot-to-dot puzzles and math games. Each activity has a symbol that tells you what kind of activity it is (see the symbol key below). Check the symbol first and then read the activity instructions carefully. You will find the stickers in the middle of the book and the answers on page 32.

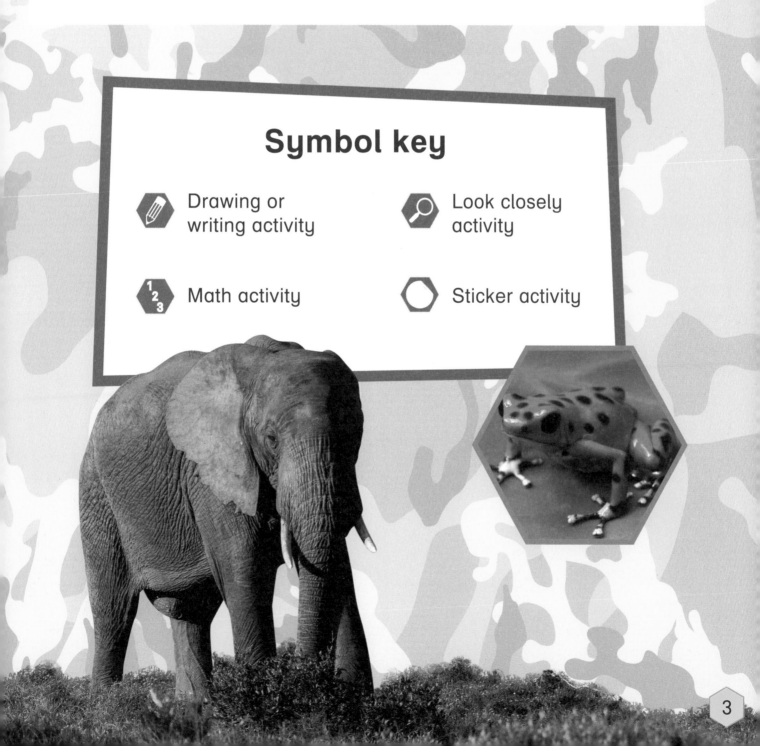

Symbol key

✏️ Drawing or writing activity

🔍 Look closely activity

1 2 3 Math activity

⬡ Sticker activity

HABITAT: DESERTS

The Sun's searing rays scorch the desert sand and rocks, creating one of the world's most extreme habitats. Deserts are not only hot, but are very dry too. The heat and dryness make it difficult for plants and animals to live, but some extreme survivors manage to make deserts their home.

Connect each desert animal to its name and add the stickers.

Camel   Rattlesnake   Tortoise   Thorny devil   Jerboa

 Can you find these desert words in the word search?

G	W	M	Z	T	S	W	C	E	C
T	W	C	A	C	T	U	S	K	E
R	G	P	I	C	H	R	I	L	O
T	O	J	T	S	C	M	M	I	A
U	B	K	S	P	S	S	P	O	S
L	I	K	A	L	A	H	A	R	I
C	R	P	N	D	H	R	S	S	S
H	F	B	D	M	A	T	H	L	P
P	L	G	M	Y	R	B	T	S	U
R	Y	M	O	M	A	B	T	I	S

CACTUS

OASIS

GOBI

SAND

KALAHARI

SAHARA

Scientists have kept a record of how much rain has fallen in 12 months in the Kalahari Desert in southern Africa. Add up all 12 months to find the total annual rainfall.

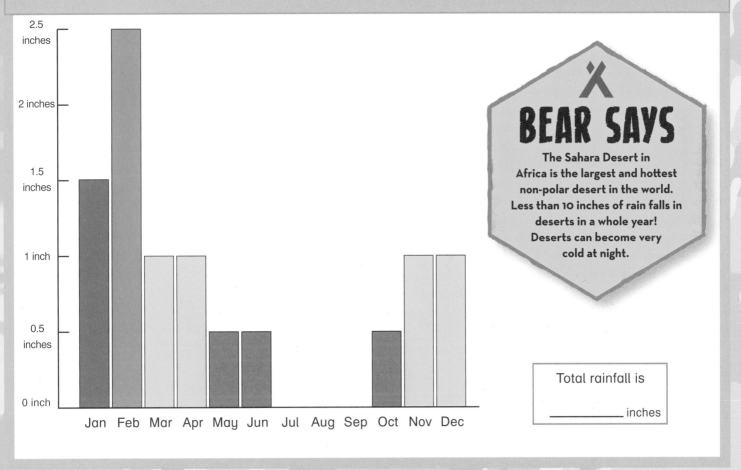

BEAR SAYS

The Sahara Desert in Africa is the largest and hottest non-polar desert in the world. Less than 10 inches of rain falls in deserts in a whole year! Deserts can become very cold at night.

Total rainfall is

_____ inches

Connect the dots to complete this picture of a fennec fox.

SUPERB SPEED

Every animal fights their own battle to survive, and some of them use speed to either chase their prey, or to escape from predators. Being fast takes a lot of energy, so animals rarely turn on the speed unless it's a life-or-death situation.

 Cheetahs can run as fast as 60 mph. Spot four differences between these two photos.

 The Australian tiger beetle can move at 5.5 mph. Draw this speedy insect using the grid as a guide.

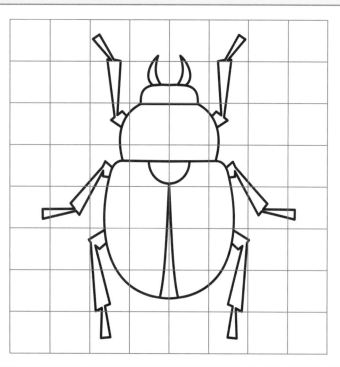 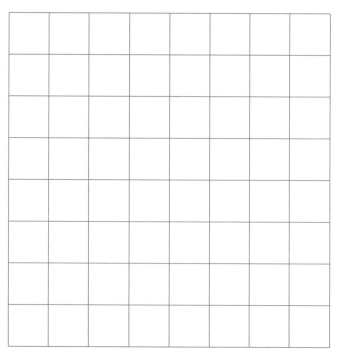

Follow the lines to match these fast-moving predators with their prey.

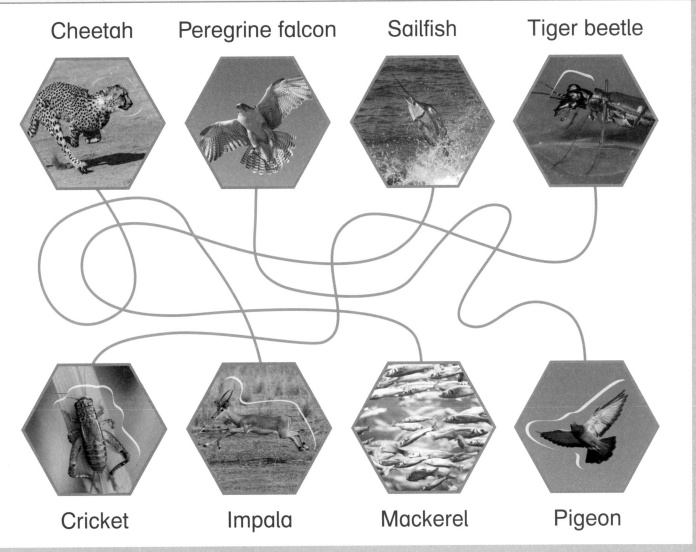

Cheetah   Peregrine falcon   Sailfish   Tiger beetle

Cricket   Impala   Mackerel   Pigeon

If a peregrine falcon swoops towards its prey at a speed of 60 mph, how far will it travel in one minute?

CLUE: There are 60 minutes in an hour.

The peregrine will travel _____ mile(s).

SMART ANIMALS

Smart animals use their brains to solve problems and to communicate, just like us. Some of the cleverest animals are monkeys and apes, but dolphins, whales, elephants, members of the dog family, and some parrots also have these useful skills.

 Some primates can use tools! Color in this picture of a capuchin using a rock to crack open a nut.

 This grey parrot is probably smart enough to find its own way through the maze to reach the fruit, but you could help it get there faster!

QUICK QUIZ

What have you learned in this section? Test your knowledge! Read the questions below and circle the correct answer.

1 Deserts get less than 10 inches of rainfall per year.

 TRUE  **FALSE**

2 The biggest non-polar desert in the world is the Gobi Desert.

TRUE **FALSE**

3 The Australian tiger beetle is very slow.

TRUE **FALSE**

4 Cheetahs can run as fast as 60 mph.

TRUE **FALSE**

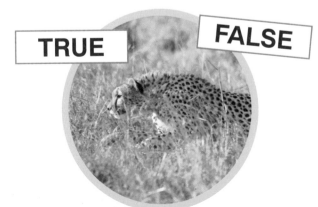

5 Elephants are unintelligent.

TRUE **FALSE**

HABITAT: POLAR REGIONS

The Arctic and the Antarctic are the Earth's icy extremes. The Arctic is the largely frozen ocean and land that surrounds the North Pole, and the Antarctic is the land around the South Pole. The animals in these places must cope with very long, cold, and dark winters. Even the summers are cool, and much of the ground and sea remains covered in ice.

 This Arctic fox is lost in a snowstorm. She can hear her cubs calling her, but she can't find them. Can you help?

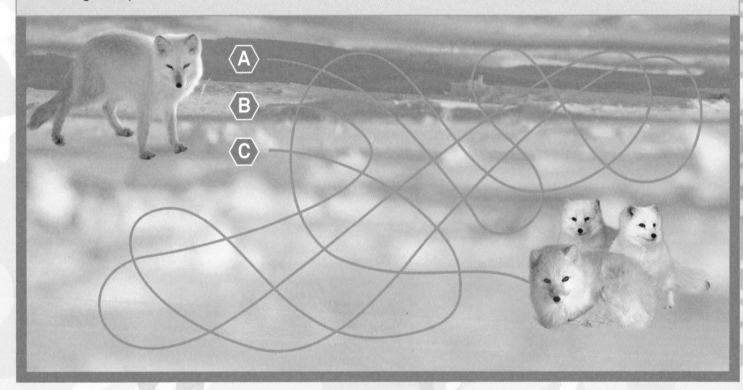

Polar animals have to deal with extreme winters and summers. At the North Pole, the Sun rarely sets in the summer. In the winter, the Sun almost never rises. If there are 18 hours of sunlight in a day, how many hours of darkness will there be? What season would it be?

CLUE: There are 24 hours in a day.

The season would be _____.

There are _____ hours of darkness.

Match the polar animal to the correct clue.

1 Walrus

2 Penguin

A I am a bird from the Antarctic, but I cannot fly.

B I am a huge white bear that hunts seals.

C I am a mammal that lives in the water and hunts in packs.

D I am a large mammal with flippers. I have two tusks.

3 Polar bear

4 Orca

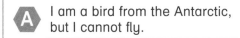

BEAR SAYS

Emperor penguins live in the Antarctic, where it is so cold that parent birds have to hold their egg on their feet to stop it from freezing!

 Unscramble the names of these animals that live in polar places.

EASL

1 _____

ENUPGIN

2 _____

RCACTI XOF

3 _____

APLRO EABR

4 _____

CRAO

5 _____

LRWUAS

6 _____

TOP MOVERS

Animals need to move to find food, shelter, or mates. While most of them have legs and can walk, there are lots of other ways of getting around.

1 2 3 Hummingbirds have the fastest wingbeat of any bird. If a hummingbird beats its wings 70 times a second, how many times will its wings beat in 10 seconds?

70 x 10 =

The hummingbird's wings will beat _____ times.

Find the stickers, then draw lines to match each animal to the way it moves.

| Jumping | Slithering | Swinging | Gliding |

Squid and octopuses move by jet propulsion, squirting water quickly away from their bodies to push themselves through the water. Squid can change their color in a second, producing shimmering waves of bright colors. Color in this squid using bright colors.

Kangaroos are champion jumpers—they can leap 26 feet in a single bound! What is a kangaroo baby called?

BEAR SAYS

Spiders that live in deserts can cartwheel across the hot ground so they don't burn their feet! Some fish can use their fins like legs to crawl through mud. These fish are called mudskippers.

A kangaroo baby is called a:

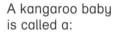 Jimmy

 Jackie

 Joey

DEFEND OR DIE

The ability to defend yourself and your young from attack is an important survival skill for all animals. The most extreme ways of keeping attackers away include deadly venoms, clever tricks, and even suits of armor!

 These two venomous vipers look similar, but there are five differences to spot.

 Draw a line between each animal and its method of defense.

 1

 2

 3

 4

A Toxic skin that can kill a predator that touches or licks it.

B Long tentacles that sting with venom.

C Strong, scaly skin too tough for a predator's jaws.

D Playing dead to fool predators.

14

QUICK QUIZ

What have you learned in this section? Test your knowledge! Read the questions below and circle the correct answer.

1 The Antarctic is the region around the North Pole.

TRUE   FALSE

2 At the North Pole the Sun never sets in the winter.

TRUE   FALSE

3 Hummingbirds have the fastest wingbeat of any bird.

TRUE   FALSE

4 Octopuses swim by wiggling their tentacles.

TRUE   FALSE

5 Opossums defend themselves with a venomous bite.

TRUE   FALSE

HABITAT: DEEP SEA

The deep ocean is one of the most incredible habitats on Earth. Light cannot pass through the water beyond a depth of about 3,280 feet, where the ocean becomes a dark, silent place. The weight of the water is too great for many animals to live, and there is no plant life in the deep seas as plants need sunlight to survive.

 Animals in the deep sea make their own light, and can glow in a range of colors. Color in this jellyfish in bright pinks, greens, and blues.

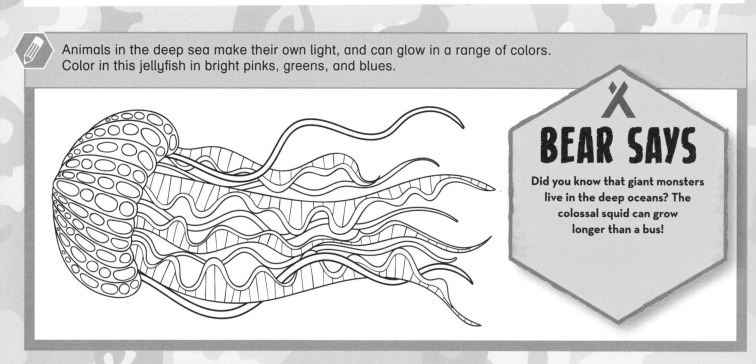

BEAR SAYS

Did you know that giant monsters live in the deep oceans? The colossal squid can grow longer than a bus!

 Complete the dot-to-dot puzzle to reveal an amazing marine monster—the angler fish.

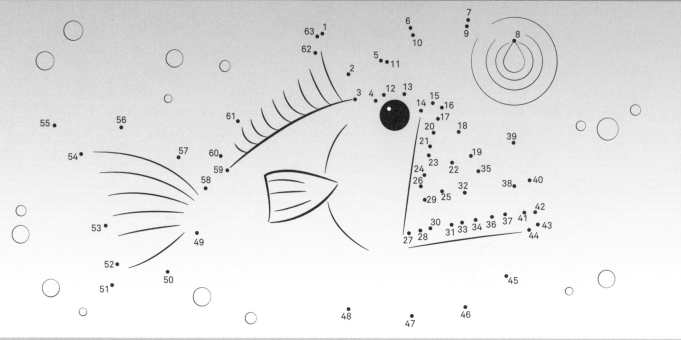

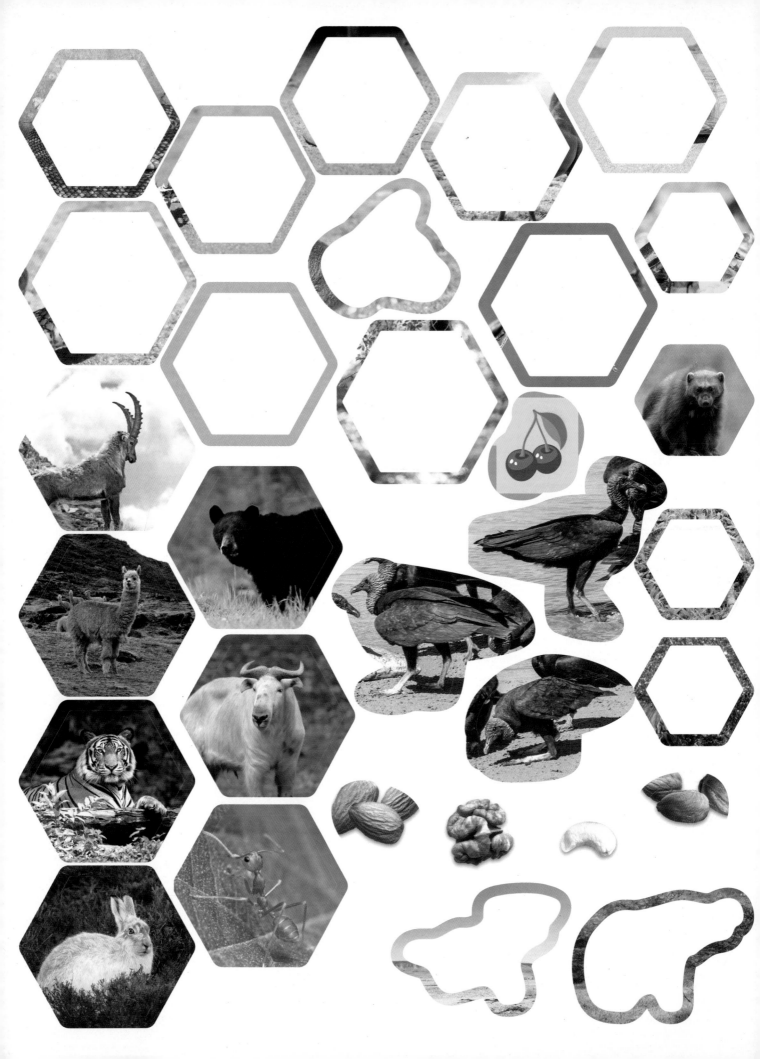

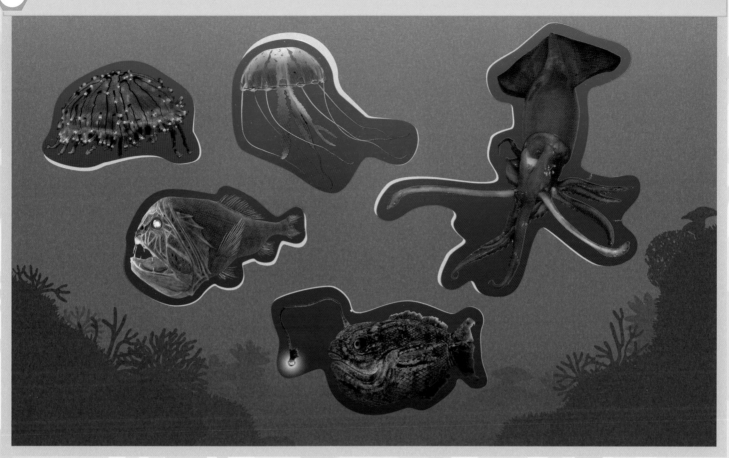

Follow the lines to discover the names of these strange deep-sea animals.

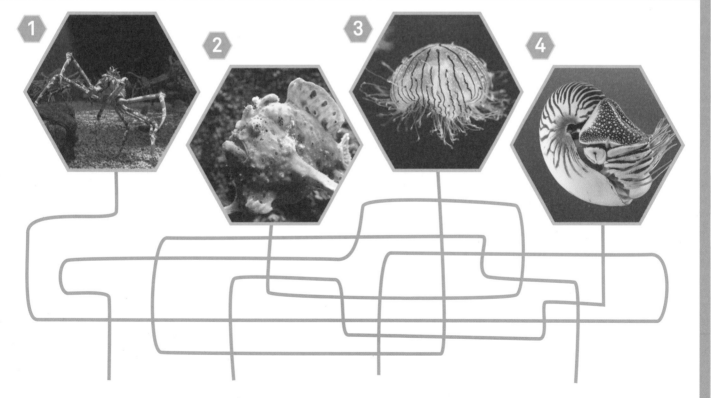

Yellow anglerfish   Nautilus   Giant spider crab   Flower hat jellyfish

SUPERB SENSES

Animals use their senses to find out about the world around them. The five major senses are sight, hearing, touch, taste, and smell, but many animals have extra special senses they can use. Some animals can sense the Earth's magnetic field, and use this to find their way. Others have an electro-sense, which means they can detect the electricity made in another animal's muscles.

 Each of these animals is missing something! Complete the drawings: draw in the spider's eyes, the hare's ears, and the moth's missing antennae.

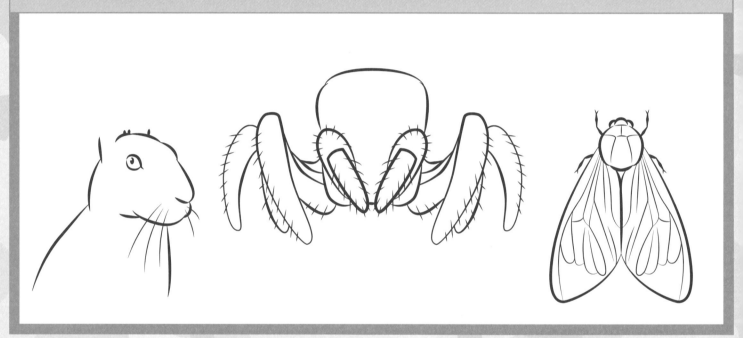

A duck-billed platypus has up to 40,000 special cells on its bill that it uses to detect the electricity produced by its prey, hiding in the mud at the bottom of the river. Put the sticker of this strange mammal on the picture.

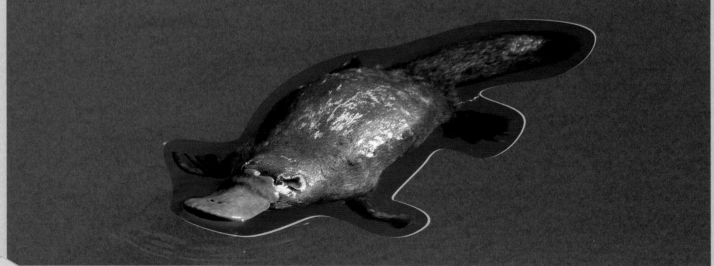

Insects see colors differently than humans; they see brighter colors than we do, but they can't see red. Follow the color key below to show how humans, bees, and butterflies see the same flower.

Human

Bee

Butterfly

Color key

1—orange   2—purple   3—green   4—yellow   5—red   6—leave white

Bats use echolocation, making high-pitched noises and listening for the echo as the sound bounces off objects in the dark. Help this bat find its way to its insect prey through the dark.

BEAR SAYS

An eagle's eyesight is 10 times better than a human's and it can see a rabbit two miles away. Moths can smell the nectar in a flower that is so far away a human couldn't even see it!

CHAMPION WEIGHTLIFTERS

Having extreme strength helps an animal in its daily battles with enemies. It also makes them tough and more likely to survive if they take a tumble or if they find themselves in a predator's jaws! These champions are among the strongest animals on Earth.

 Some small animals have surprising strength! Draw a line between the animal and its super-strength ability.

Peacock mantis shrimp

1

Coconut crab

2

Rhinoceros beetle

3

Wolverine

4

A Can carry 850 times its own weight on its armored back!

B Has a very strong bite for its size!

C Can punch through the glass of an aquarium tank.

D Has the strongest pinch and can break a coconut open with its claws.

 Pangolins are armor plated. Continue the pattern of bony scales on its back. Add the sticker of a pangolin rolled up to protect itself.

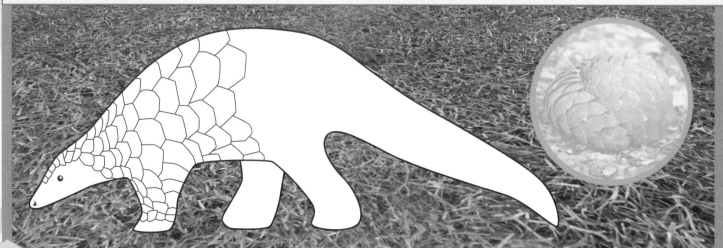

QUICK QUIZ

What have you learned in this section? Test your knowledge! Read the questions below and circle the correct answer.

1 Only plants exist in the deep ocean.

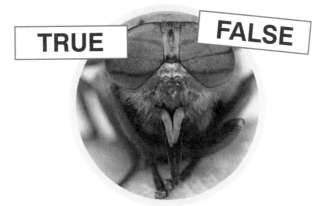

TRUE **FALSE**

2 Animals in the deep sea often create their own light.

TRUE **FALSE**

3 Insects see colors differently than humans.

TRUE **FALSE**

4 Duck-billed platypuses can sense electricity in their prey.

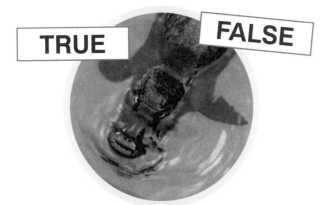

TRUE **FALSE**

5 Peacock mantis shrimp aren't very strong.

TRUE **FALSE**

HABITAT: TOWNS AND CITIES

Many animals and plants have moved out of the natural world and into an extreme urban habitat of bricks, glass, and concrete. They have made homes in the most unusual places, and found food wherever humans live, in towns and cities all over the world.

 This scene of langur monkeys in an Indian town has an intruder! Can you name the animal that doesn't belong here?

 This cheeky chipmunk has been collecting nuts for his store, where he keeps food for the winter. Place the stickers, then add up all the nuts and write the total in the box.

Almonds   Walnuts   Brazil nuts

Cashews

Almonds: _____	nuts
Walnuts: _____	nuts
Brazil nuts: _____	nuts
Cashews: _____	nuts
Total: _____	nuts

Complete the dot-to-dot puzzle to reveal a big cat that sneaks into towns to hunt, then draw in lots of spots! Unscramble the letters to reveal its name.

EARLDOP

_____

 Rats live happily in towns and cities, as there is always plenty of food. They can even make their home in sewers. Help this rat find its way through the underground pipes.

BEAR SAYS

The word "urban" is used to describe the habitats where many humans live together, such as towns and cities. Many animals now have urban habitats too.

SWARMS

There is strength in numbers, and that's why some animals gather in huge groups. By sticking together, a group can look out for predators and travel more safely. When some groups go on the move, they create swarms, which can have the power to strip trees and crops of leaves, fruit, and seeds.

 These mackerel are swimming in a huge shoal to make it difficult for predators to pick out one fish from the swirling, silvery cloud. They all look alike, but one is different. Circle the odd fish out.

 100 locusts eat 10 plants in one hour. How many plants will they eat in two hours?

100 locusts will eat _____ plants in two hours.

Flocks of flamingos gather together at water holes for protection, and to meet mates. Sometimes they dance! Color in these two dancing flamingos using lots of pinks and greens.

BEAR SAYS

Periodical cicadas are insects that swarm every 17 years or so. Millions of them emerge from the ground over just a few days and nights.

Unscramble the group names below, then draw a line to match the animals to their group.

1

Penguins

2

Bees

3

Mackerel

4

Wolves

MRWSA

HOASL

ACPK

NYOLCO

_____ _____ _____ _____

TOTALLY GROSS

Being big, strong, or speedy isn't enough for some animals. Instead, they use the gross factor to achieve success, scavenging for leftovers. These bizarre animals are just a few examples of the many creatures that go to extreme lengths to stay alive!

 Use your stickers to complete this scene of scavenging vultures feeding on a carcass (the body of a dead animal).

 Unscramble the letters and draw a line to connect these creatures to their names. These are all scavengers that feed on waste, or the bodies of animals that predators have left behind.

WROC   GTOAMG   BRCA   TRA

_____   _____   _____   _____

QUICK QUIZ

 What have you learned in this section? Test your knowledge! Read the questions below and circle the correct answer.

1 Rats sometimes live in sewers.

TRUE   **FALSE**

2 Leopards never enter cities as they are scared of humans.

TRUE   **FALSE**

3 It is dangerous for animals to travel in groups and they rarely do.

TRUE   **FALSE**

4 A group of bees is called a gang.

TRUE   **FALSE**

5 Vultures will eat dead bodies that they have found.

TRUE   **FALSE**

HABITAT: MOUNTAINS

Mountains are extreme habitats that get a lot of sunlight on long summer days, but can turn extremely cold at night and are often snowcapped in the winter. Plants struggle to survive on windswept peaks, and slippery rocks are challenging to all except the most sure-footed animals.

 Write true or false under each statement about this beautiful snow leopard.

1 It has thick fur to keep it warm.

2 It's a member of the dog family.

3 Its white fur and spots help it stay hidden.

4 It eats berries, fruits, and seeds.

 This Alpine ibex and takin are making their way along the craggy ledges of a mountain. Follow the lines to find out who will reach the summit.

Draw some mountain animals onto this mountainside habitat.

Find the right stickers and use the first letter from each animal's name to create a word that means an animal's home.

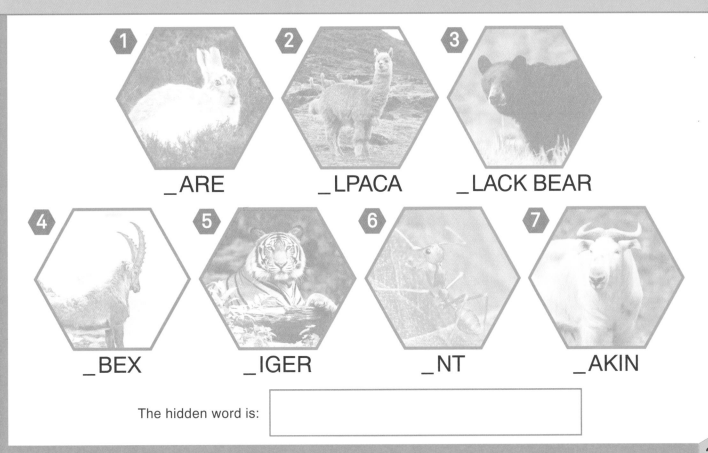

1 _ARE

2 _LPACA

3 _LACK BEAR

4 _BEX

5 _IGER

6 _NT

7 _AKIN

The hidden word is:

ANIMAL GIANTS

At the time of the dinosaurs, many enormous animals lived in the sea and on land. Today, there are fewer giants, but the blue whale is still the biggest animal that has ever lived. Being big has its advantages: larger animals can keep warm more easily, they can fight their enemies better, and they can often reach food that smaller animals cannot.

 Follow the lines to match each giant animal to its size.

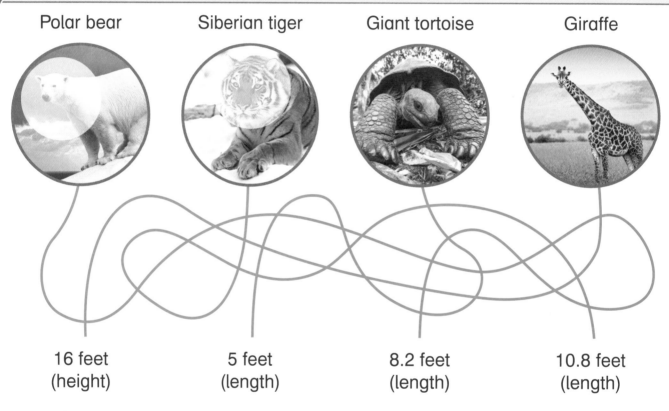

Polar bear   Siberian tiger   Giant tortoise   Giraffe

16 feet
(height)

5 feet
(length)

8.2 feet
(length)

10.8 feet
(length)

 Circle all the words that mean "big."

ENORMOUS   MINUSCULE   MINI

COLOSSAL

MODEST   HUGE   MEAGER   MINIATURE

GIGANTIC   PALTRY   VAST   DIMINUTIVE

IMMENSE

LITTLE   SLIGHT   MICROSCOPIC

Complete the dot-to-dot puzzle to reveal the largest land mammal.

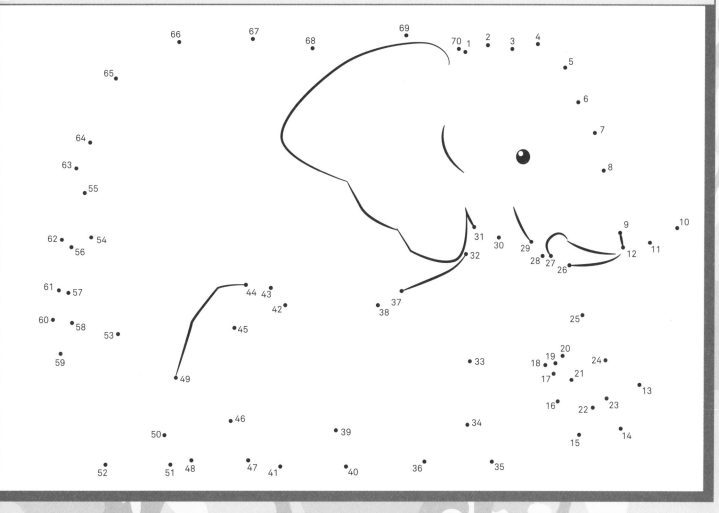

Find and add the sticker for each record breaker!

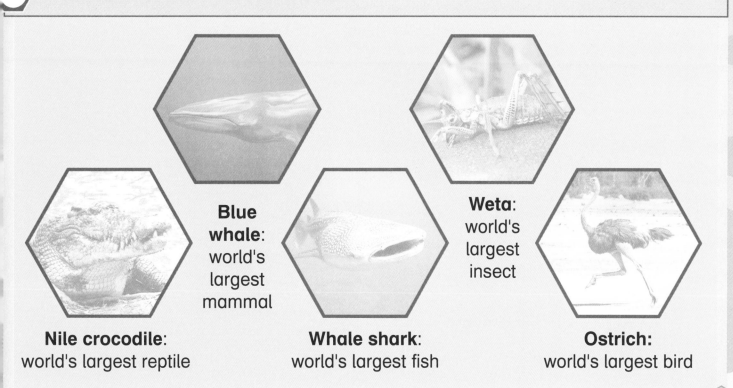

Blue whale: world's largest mammal

Weta: world's largest insect

Nile crocodile: world's largest reptile

Whale shark: world's largest fish

Ostrich: world's largest bird

ANSWERS

Page 4
Connect the animals
1. Rattlesnake
2. Thorny devil
3. Jerboa
4. Camel
5. Tortoise

Desert word search

Page 5
Rainfall graph
Total rainfall is 9.5 inches.

Fennec dot-to-dot

Page 6
Cheetah spot the difference

Page 7
Follow the lines
Cheetah—Impala
Peregrine falcon—Pigeon
Sailfish—Mackerel
Tiger beetle—Cricket

Peregrine distance
The peregrine will travel 1 mile.

Page 8
Parrot maze

Page 9
Quick quiz
1. True; 2. False; 3. False;
4. True; 5. False.

Page 10
Navigate the snowstorm
Route B is the correct path to
the cubs.

Polar sunlight
The season would be summer.
There are 6 hours of darkness.

Page 11
Match the polar animal
1. D; 2. A; 3. B; 4. C

Unscramble the names
1. SEAL
2. PENGUIN
3. ARCTIC FOX
4. POLAR BEAR
5. ORCA
6. WALRUS

Page 12
Hummingbird wings
The hummingbird's wings will beat
700 times.

Animal movers
1. Slithering
2. Gliding
3. Swinging
4. Jumping

Page 13
Baby kangaroo
A kangaroo baby is called a joey

Page 14
Viper spot the difference

Animal defense
1. D; 2. A; 3. B; 4. C

Page 15
Quick quiz
1. False; 2. False; 3. True;
4. False; 5. False.

Page 16
Angler fish dot-to-dot

Page 17
Follow the lines
1. Giant spider crab
2. Yellow anglerfish
3. Flower hat jellyfish
4. Nautilus

Page 19
Bat maze

Page 20
Strong animals
1. C, 2. D, 3. A. 4. B.

Page 21
Quick quiz
1. False
2. True
3. True
4. True
5. False

Page 22
Monkey intruder
The chimpanzee is the animal that
doesn't belong in the image.

Cheeky chipmunk
3 Almonds
2 Walnuts
3 Brazil nuts
6 Cashews
14 nuts in total.

Page 23
Dot-to-dot
Answer: Leopard
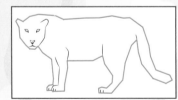

Pipe maze

Page 24
Odd mackerel out
One of the fish (at the top) is pink.

Locust feast
100 locusts will eat 20 plants in
two hours.

Page 25
Animal group scramble
1. Penguins—Colony
2. Bees—Swarm
3. Mackerel—Shoal
4. Wolves—Pack

Page 26
Scavenger scramble
1. Maggot
2. Rat
3. Crow
4. Crab

Page 27
Quick quiz
1. True; 2. False; 3. False;
4. False; 5. True.

Page 28
Snow leopard facts
1. True
2. False
3. True
4. False

Summit race

Page 29
Animal home
H A B I T A T

Page 30
Follow the lines
Polar bear—8.2 feet long
Siberian tiger—10.8 feet long
Giant tortoise—5 feet long
Giraffe—16 feet high

Big Words
enormous, colossal, huge, gigantic,
vast, immense

Page 31
Dot-to-dot